Giuseppe Agrillo

Weather predictions for a volcanic ash cloud tracking system

AF154043

Giuseppe Agrillo

Weather predictions for a volcanic ash cloud tracking system

A context of High Spatial Resolutions

LAP LAMBERT Academic Publishing

Impressum / Imprint

Bibliografische Information der Deutschen Nationalbibliothek: Die Deutsche Nationalbibliothek verzeichnet diese Publikation in der Deutschen Nationalbibliografie; detaillierte bibliografische Daten sind im Internet über http://dnb.d-nb.de abrufbar.

Alle in diesem Buch genannten Marken und Produktnamen unterliegen warenzeichen-, marken- oder patentrechtlichem Schutz bzw. sind Warenzeichen oder eingetragene Warenzeichen der jeweiligen Inhaber. Die Wiedergabe von Marken, Produktnamen, Gebrauchsnamen, Handelsnamen, Warenbezeichnungen u.s.w. in diesem Werk berechtigt auch ohne besondere Kennzeichnung nicht zu der Annahme, dass solche Namen im Sinne der Warenzeichen- und Markenschutzgesetzgebung als frei zu betrachten wären und daher von jedermann benutzt werden dürften.

Bibliographic information published by the Deutsche Nationalbibliothek: The Deutsche Nationalbibliothek lists this publication in the Deutsche Nationalbibliografie; detailed bibliographic data are available in the Internet at http://dnb.d-nb.de.

Any brand names and product names mentioned in this book are subject to trademark, brand or patent protection and are trademarks or registered trademarks of their respective holders. The use of brand names, product names, common names, trade names, product descriptions etc. even without a particular marking in this work is in no way to be construed to mean that such names may be regarded as unrestricted in respect of trademark and brand protection legislation and could thus be used by anyone.

Coverbild / Cover image: www.ingimage.com

Verlag / Publisher:
LAP LAMBERT Academic Publishing
ist ein Imprint der / is a trademark of
OmniScriptum GmbH & Co. KG
Heinrich-Böcking-Str. 6-8, 66121 Saarbrücken, Deutschland / Germany
Email: info@lap-publishing.com

Herstellung: siehe letzte Seite /
Printed at: see last page
ISBN: 978-3-659-68177-6

Zugl. / Approved by: Naples (Italy), University of Naples "Federico II", Diss., 2014

"Avoid, Avoid, AVOID!"

The procedure recommended
by ICAO on vulcanic ash clouds

Contents

List of Figures

1

Abstract

The CCMMMA (Centro Campano per il Monitoraggio e la Modellistica Marina ed Atmosferica) is the marine weather center of the University of Naples "Parthenope" and its purpose is expose multi-disciplinary services for the environmental monitoring, produce weather, marine and air quality forecast for Europe, Italy and Campania region (southern Italy) at as high as possible spatial resolution. The weather data of CCMMMA has been used to customize a prototype system focused on volcanoes of southern Italy that allows the on-demand forecast of ash cloud dispersion. The aim is to deploy a workflow of coupled models embedded into an advanced computing system, able to provide on-demand hourly simulations oriented to the dispersion of ash clouds from volcanic eruptions. The daily weather output of the Weather and Research Forecast (WRF) model was coupled with the CALPUFF modeling system to obtain a hourly forecast dispersion of volcanic ash clouds in Southern Italy. The whole process is deployed through a bash-scripting wrapper that allows the execution of the workflow in a few minutes; this process could be useful for early-warning system and to improve the studies about flight safety, environment protection and climate change. In this work some case studies, together with the validation of results, will be shown.

Introduction

A volcanic ash eruption is a dangerous condition that have both short- and long-term consequences over human activities. Using a monitoring network it is possible to observe an eruption and define a rescue plain to improve the volcanic ash risk assessment. The ICAO distinguishs between hazards from volcanic ash for aircraft safety and for the longer term safety. A volcanic ash reduces visibility, rendering windscreens partially or completely opaque; onboard sensors may result unreliable; particles can fuse in the turbine reducing its efficiency, or even stop it; a contamination in the cabin requires the use of oxygen masks for the crew; the flight crew maneuvering may potentially conflict with other aircrafts in the vicinity; deposits of volcanic ash on a runway results in the degradation of braking performance, especially if the volcanic ash is wet; in extreme cases, this can lead to runway closure.

The eruptions from stratovulcanoes can be explosive with emission of both lava and pyroclastic flow, evolving in a high temperature column composed by a heterogeneous mixture of components with irregular shape and very hard material. The plume could achieve altitudes of a few kilometers from sea level and micrometric particles could be dispersed in the atmosphere thousand of kilometers away from the vulcano. Clouds may be so dense to obscure the sunlight and reduce the flight visibility. Closer to the volcano, dense ashfall can smother crops, leaving land unproductive for a long time afterwards, and in some cases can cause structural collapse of buildings, thereby endangering life and infrastructures. In high population density areas the infrastructures may be damaged, the human health risks increase and the population is exposed to death risk. Airborne ash and SO_2 are also known to cause respiratory problems for both humans and livestocks, even at relatively low concentrations.

A volcanic eruption with emission of ashes makes the flight, landing and takeoff conditions very hazardous and the aircrafts should never fly within an ash cloud. Also, when the kinetic energy and the temperature decrease, ashes fall on the ground at high velocity and everything will be covered by an ash layer. The eruption of Eyjafjallajokull in May 2010 demonstrated the huge economic costs that can follow from airspace closure, and the widespread disruption to freight-related services worldwide, highlighting how critical is to know and constrain the hazard in space and time.

An aircraft can avoid to fly into ash clouds thanks to presence of early-warning systems involved to forecast of ash cloud dispersion in the atmosphere after an eruption. A monitoring system is required in order to issue timely warnings to avoid the hazard conditions. Further, the accurate observation of volcanoes is important activity to obtain a recognition and characterization of the ash emissions because these informations can be used to predict the evolution of plume and the transport of ash into atmosphere.

The Mediterranean Basin, is characterized by a complex schema of tectonic dynamics because Eurasian plate and African plate are connected under the Mediterranean sea. Italy is crossed longitudinally from South to North by a deformation band, all along the Apennines. In the middle of the Mediterranean Basin, in southern Italy, there is a depression in the sea floor located between a subduction zone and an associated volcanic arc. In this area we have few of the most famous and potentialy dangerous active volcanoes in the world, such as the Stromboli and Vulcano in the Aeolian Islands and Mt. Etna in Sicily, close to the urban area of Catania, and other vulcanic areas classified as "quiescent" such as those in the Campania region near Naples: Campi Flegrei and Vesuvius. The presence of these volcanos motivates our work: develop a prototype system that can use weather data (e.g. wind, temperature) at a high spatial resolution centered on southern Italy volcanoes: Stromboli, Etna and Vesuvius.

Our system couples the WRF model (one of the most advanced mesoscale weather model) with the CALMET-CALPUFF system (a local- to long-range transport model) in order to timely execute an on-demand workflow using the HPC-GPU BlackJeans cluster of the University of Naples "Parthenope".

Our coupled modeling system called ACT (Ash Cloud Tracking system) and composed by WRF, CALMET and CALPUFF models, can choose the best weather data available and produce hourly 3D maps of ash dispersion. Also, we developed a KML Engine module: a software to display the output using the standard tools of Google Earth.

To facilitate the system portability, the whole process is deployed through a bash-scripting wrapper which allows a workflow execution in a few minutes, as for an early-warning tool, that could be helpful to improve flight safety, environment protection and climate change studies.

In this work we show the preliminary results obtained using the ACT system and how we can match it with reports of INGV (Istituto Nazionale di Geofisica e Vulcanologia) and NASA (the National Aeronautics and Space Administration of the United States of America) satellite images that show the eruption of Stromboli and Mt. Etna.

This thesis is divided in the following chapters: in Chapter 1, I provide an overview about volcanic ash and their impacts on aviation (aircrafts and airports) and the official safety rules. Chapter 2 shows the state-of-the-art about prediction of dispersion processes of an ash cloud in the atmosphere. In Chapter 3, I describe the components of our system and the interface that allow to produce an ash dispersion map. Chapter 4 illustrates the applications of our system to Stromboli and Mt. Etna. Finally, Chapter 5 shows the conclusions and how to improve and develop this work in the future.

1 Impact of Volcanic Ash Clouds on aviation

Volcanic ash clouds are composed by pulverized rocks, sharp-edged, hard glass particles. As reported in paper [1] the particles are abrasive and have a melting temperature below the operating temperature of modern turbine engines at cruise velocity because they are largely composed by siliceous materials. Within a volcanic ash cloud we can also have a mixture of gases, i.e. sulfur dioxide (SO_2 which, when combined with water, forms sulphuric acid), chlorine (which, when combined with water, forms hydrochloric acid) and other chemicals. According to ICAO [4] all these components are corrosive for the aircraft and dangerous for human health. So it is easy to understand that volcanic ash in the atmosphere may be a serious danger in the management of an airport and for flying aircrafts, and aircrafts should always avoid it.

1.1 Volcanic ash risk assessment

In a volcanic eruption we have a wide range of hazards that we need to manage. Knowledge of the characteristics of a volcanic eruption can improve the volcanic risk assessment and save many lives. However, this task is difficult to be made because the characteristics of a volcanic eruption are extremely variable. An eruption has a spatial range from a few kilometers to thousands, and the time range can change from few minutes to hours. If a volcanic eruption is also characterized by the emission of volcanic ash, the ashes can be transported by the wind and, when the ashes fall, they can cover large areas damaging infrastructures and have a high socio-economic impact.

Figure 1.1: 30 October 2002, Eruption of Mt. Etna with fomation of plume and ash clouds (source: NASA, ISS005-E-19024)

From the point of view of air traffic management, a volcanic ash cloud can be very dense, particles can damage the aircraft and its engines and they can also reduce the flight visibility and the ground visibility. When ashes fall they reduce the usability of paved surfaces, including airport runways. Ash cleanup from roads and airports is commonly necessary, but the large volume makes it a logistic challenge.

As reported in the paper [1] an eruption of volcanic ashes can severely impact on regional transportation networks and large areas, including airports and aircraft, can be unreachable from hours to days.

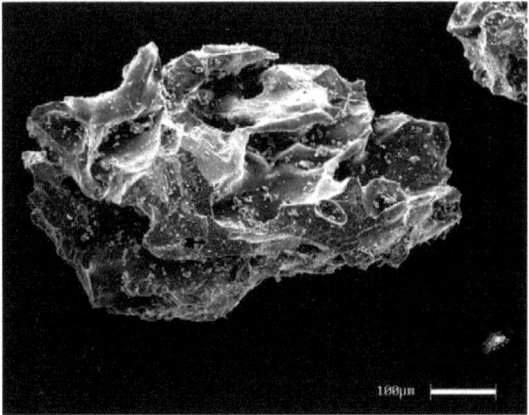

Figure 1.2: Sample of volcanic ash particle (source: [1]).

Figure 1.3: Aircraft and runways covered by a volcanic ash layer

The presence of an airport in a volcanic area is more common than we might think. In these cases the main hazard for an airport is the ash fall, because few millimeters are sufficient to close its runways. According to [5], the airports impacted by volcanic activity

13

from 1944 to 2006 are 101 in 28 countries were affected on 171 occasions by eruptions of 46 volcanoes. Among these volcanoes which caused great impact there is Mt. Etna in Italy (37.7550° N, 14.9950° E).

The eruptions of Etna are characterized by Strombolian activity: effusion of lava flows and ash emissions. In the last hundred years it made a succession of low-energy explosive eruptions and lava effusions. These eruptions, lasted several days or even years, repeatedly damaged the urban areas due to the accumulation of ash. Etna ash emissions are not hazardous to human life, but they cause disruption to the transport networks, have a severe socio-economic impact, and, in the case of heavy exposure, even respiratory diseases. The fallout of ash caused significant damage to agriculture, severe disruption to air traffic management to the airport of Catania-Fontanarossa-Sigonella and Reggio Calabria.

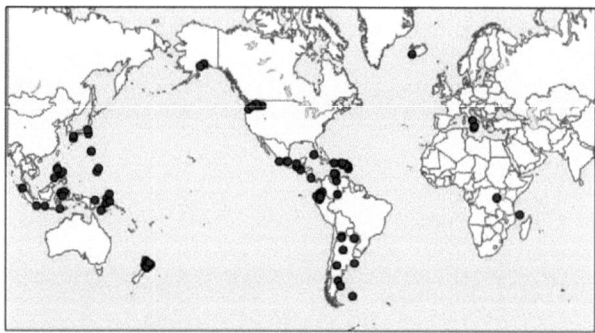

Figure 1.4: Locations of airports impacted by volcanic eruption from 1944 to 2006 as reported by [2]).

Today is still alive the memory of the Eyjafjallajökull eruption in April–May 2010, which caused the closure of the European and North Atlantic airspace. European airports were closed for many days and airlines followed the ICAO rule of "zero tolerance" about the ash clouds presence in airspace, according to [1] almost 95,000 flights were canceled in the European airspace, causing large economic losses to airline companies, to the insurance industry and to companies depending on European markets.

In order to mitigate hazards, volcanologists and atmospheric scientists developed advanced instruments, analysis tools and softwares with the aim to solve numerical models for the observation and prediction of volcanic ash behavior.

The impacts of volcanic activity on an airport can be mitigated if the flight and ground operations are timely planned. The success of the operations is improved if the air traffic management system receives a forewarning of volcanic hazards.

I show in this thesis that this target can be achieved using a coupled system of models, providing weather information at high spatial and time resolution to obtain an estimation of wind fields and to calculate the transport and the dispersion of ash clouds for a specific area (more details in Chapter 2). The models need computing resources to numerically solve the physical and chemical equations which describe the transport and the dispersion of ashes and predict their spatial and time evolution. If we integrate these systems within an early-warning monitoring network, it is possible to save lives and reduce the socio-economic impacts of a volcanic eruption with ash fall.

With this approach we can provide a useful system available to public and private entities to support the operations and the decisions which must be taken when an event occurs with the emission of volcanic ash. The air traffic controllers can timely dispose the cover of parked aircraft and other equipments; it can reduce the closure time, optimizing runway usage before and during eruptive events, and modifying the take-off routes to avoid ash in nearby airspace [5].

1.2 Flight Safety and Volcanic Ash

The flight through a volcanic ash cloud or landing/takeoff phases by an airport with runways covered by ashes is a hazard to flight safety from many points of view. The study of volcanic ash clouds is a multidisciplinary field where volcanologists, atmospheric scientists and engineers work together sharing their knowledge [6]. The following short-list provides some examples of what we can expect when there is a volcanic eruption with the emission of volcanic ash cloud and what may be the consequences:

- the malfunction or failure of one or more engines, leading also failures of electrical, pneumatic and hydraulic systems;

- the blockage of sensors, resulting in unreliable indications and erroneous warnings;

- windscreens rendered partially or completely opaque;

- contamination of cabin air, requiring the use of oxygen masks by the crew;

- erosion of external aircraft components;

- fallout of volcanic ash on runways, resulting in a degradation of braking performance, especially if the volcanic ash is wet: this can lead to runway closure.

The timely availability of reliable information about volcanic ash is essential to mitigate the safety risk of an aircraft encountering volcanic ash and for an airport to apply the planned procedures to protect its runways and parked aircraft. In 1987, ICAO started the process, designating nine regional Volcanic Ash Advisory Centers (VAACs): Anchorage, Washington, Buenos Aires, Montreal, London, Toulouse, Tokyo, Darwin and Wellington. The VAAC's mission is to ensure that volcanic ash cloud hazards are identified when they occur, producing advisories to the aviation community; VAACs are also an interface between volcano observatories, meteorological watch offices and air traffic control centers [6].

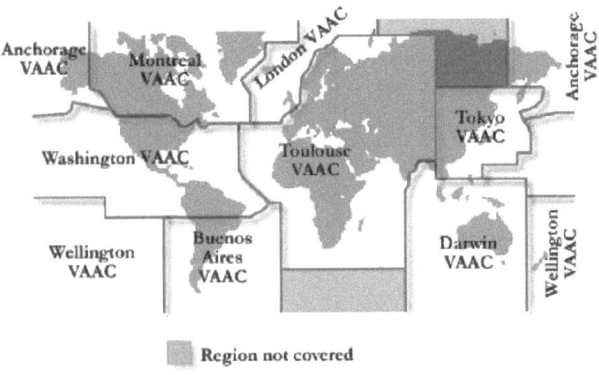

Region not covered

Figure 1.5: VAACs airspace responsibility (source: UK MetOffice)

The volcano observatories and meteorological offices collect data from their monitoring systems by ground-base, airborne and satellite-based remote-sensing systems [4].

In this scenario, the purpose of this PhD project was the set-up and the optimization of a weather forecast service at high spatial and time resolution, for a volcanic ash cloud tracking system to solve on-demand numerical models using a distributed computing system (Figure 3.2). I also contributed to the creation of a weather forecast system by developing the software tools needed to manage all steps ensuring the execution of numerical models about weather, sea state and dispersion of pollutants. To provide a forecast of volcanic ash cloud useful to aviation users, like flight controllers, public and private organizations which manage the territory and the airspace, it is necessary to deploy a fast and reliable system based on coupling Eulerian-Lagrangian environmental numerical models. In this work I used a weather prediction dataset at high spatial and time resolution to initialize a well known dispersion model that reproduced a volcanic ash cloud distribution for the Stromboli and Etna volcanoes, which were recently characterized by intensive volcanic activity. With this contribution I checked the possibility of using updated meteorological dataset at high spatial and temporal resolution to improve:

- the forecast procedures of the dispersion of volcanic ash in the atmosphere;

- the management of air traffic, reducing the risk for airplanes to encounter a volcanic ash cloud.

Moreover this system allows the knowledge of dispersion of volcanic ash even when there were poor visibility conditions, as it might occur at night. This strategy has a degree of uncertainty because the simulation depends on eruption parameters that are difficult to estimate, such temperature and emission velocity and radius of the vent.

1.3 Volcanic ash: perspective in flight

Volcanic ash particles are dangerous to engines of aircrafts because they commonly have a glassy ground-mass that could change its material properties upon warming. When the rock particles, which composing the volcanic ash clouds, are exposed to high temperatures into aircraft engines they change from solid to liquid state because the melting temperature of rocks is close to the operational temperature of engines.

The flight accidents after encountering volcanic ash clouds have been carefully analyzed, to find the critical threshold values, but these values could not easily be defined and no conclusive results have been achieved. However, it could be established that the danger depends on several parameters, such as ash mass concentration, the time spent by the aircraft in an ash layer, the engine type and the power setting. If the mass concentrations are between 0.5–1.0 mg/m^3, it is very hard for the pilots to distinguish the volcanic ash from common clouds. The visual detectability of volcanic ash depends on many parameters, such as:

- the size, the brightness;

- the colors contrast between the airborne volcanic ash and the background;

- the illumination;

- the particle size distribution and mass concentration;

- the wavelength-dependent light scattering and absorption by the ash;

- the human perception.

In April 2010 the eruption of Iceland's Eyjafjallajökull volcano (63.63N, 19.62W, 1666 m asl) caused the most extensive restriction to the airspace over Europe. The explosive eruption and the meteorological conditions led to a fast transport of volcanic ash to central Europe. The eruption of the Eyjafjallajökull volcano lasted 39 days, which far exceeded the duration of any explosive eruption phase in Iceland in the past 30 years. According to [7] the ash layer in atmosphere had heavily changed the aerosol optical depth (AOD 0.7–1.2 at 500 nm) on Europe and peak ash mass concentrations retrieved from the Lidar measurements in Leipzig and Munich showed values of 1.0 and 1.1 mg/m^3, with an uncertainty range of 0.65–1.8 mg/m^3. To prevent a scenario similar to the "Eyjafjalla-jökull ash crisis" in the future, we need reliable tools to predict and to identify regions free of dangerous ash loads; this can be achieved using an early-warning system based on numerical model solutions that are independent from real visibility conditions.

1.4 The Atmosphere of Southern Italy

Within the CCMMMA (Centro Campano per il Monitoraggio e la Modellistica Marina ed Atmosferica) we can use the latest and powerful technologies to improve both the monitoring stations network and the computing resources needed for environmental modeling, such as weather and air quality forecast, develop an operational modeling chain, couple numerical models, and ensure daily operations into distributed computing environment. To achieve this target, the CCMMMA used its own computing resources: the HPC-GPU BlackJeans cluster (Figure 3.2). In this paragraph I described the hardware and software system that allowed to produce daily forecasts for both weather and air quality (AQ).

This target was achieved by developing a workflow based on coupling the simulations of WRF (Weather and Research Forecast) with CHIMERE (a chemistry-transport model developed at the French institute IPSL/LMD) and the AODEM module [?]. The WRF, CHIMERE and AODEM models are coupled into an operational chain, to calculate the daily forecast of Aerosol Optical Depth (AOD). The use of an AOD model is important because, particles absorb and scatter the solar radiation, influencing the energy balance and inducesing physical and chemical feedbacks on atmospheric dynamics.

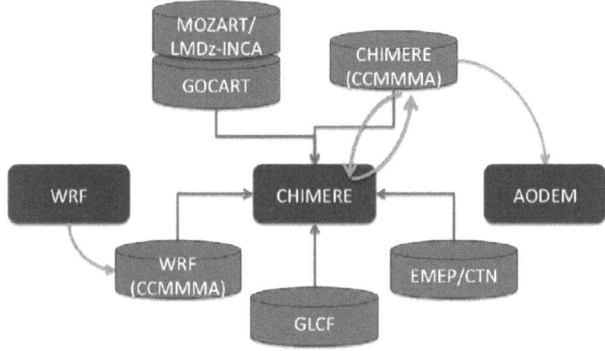

Figure 1.6: WCA block diagram and used dataset

The models chain WRF-CHIMERE-AODEM (WCA) is a workflow of numerical models, running on several nested domains for Europe, Italy and southern Italy. The execution process, comprising earth data display applications, produces maps for PM_{10} (particulate matter with an aerodynamic diameter less than 10 micrometers), Black Carbon (BC), DUST and SALT from CHIMERE data output at a spatial resolution of $0.06°$ for the innermost domain covering southern Italy. The WCA workflow runs every day on the HPC-GPU BlackJeans cluster and produces a forecast for the next 72 hours for meteorological variables and atmospheric pollutants such as temperature, rainfall, ozone, BC, PM_{10} (particulate matter with an aerodynamic diameter less than 10 micrometers) and $PM_{2.5}$ (particulate matter with an aerodynamic diameter less than 2.5 micrometers); it also produces AOD data at 550 nm wavelength. The time series trends are shown for every given station point.

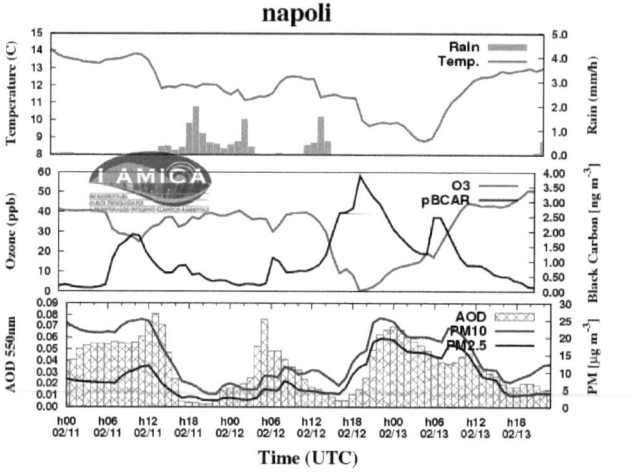

Figure 1.7: Example of a time series for station of Naples using WAC

My contribution, as shown in this thesis, is the development of a shell scripting wrapper around the WRF and CHIMERE models, and the further development of the AODEM module into the daily forecasts of CCMMMA. Our results allowed the comparison of temperature, rainfall, ozone, BC, PM_{10} and the AOD with the data collected by air quality monitoring stations. As a next step it is reasonable to propose the integration of these models with a module that takes into account the feedback contribution of volcanic ash emitted in the atmosphere as a result of eruptive events on the trend of the AOD.

2 Prediction of atmospheric dynamics

Today there are several numerical codes which have been released and officially used to describe the dispersion and deposition fluxes of volcanic ash under the action of meteorological conditions. These software are typically used to provide a qualitative description of the ash cloud transport in atmosphere; a single forecasting simulation is often the only approach available to describe an on-going, or future scenario, event. In this chapter I will show in detail which are the most advanced numerical models needed to make stable and reliable processes. With modeling approach I can track and predict dispersion of volcanic ash cloud with high spatial and time resolution [8].

2.1 Prediction and tracking of volcanic ash clouds

Volcanic ash clouds are commonly produced by volcanic eruptions. The energy of the eruption can lead to the formation of a plume: a column of ashes which rise in the atmosphere from the volcano vent (Figure 1.1).

The height reached by the plume depends on physical and thermodynamic parameters and its diffusion in the atmosphere depends on the emitted mass and the fine mass fraction responds at the buoyancy force on time scales from hours to days [6].

In our simulations we need to consider these chemical and physical properties of the atmospheric dynamics, to reduce errors, develop a realistic scenario for these processes and to mitigate the volcanic ash cloud risk [6]. Satellite remote sensing cannot provide all information on ash cloud properties and cannot provide a forecast for the ash cloud trajectory.

To forecast the ash cloud motion, volcanic ash transport and dispersion models (VATD) have been developed. The main users of these kind of softwares are: volcano observatories, meteorological offices, air traffic controllers or public and private agencies involved into the assessment of volcanic ash risk (e.g. VAACs) [6].

Examples of VATD are: the HYSPLIT model, developed for tracking atmospheric applications like ash, nuclear and chemical pollution; the PUFF model used by The Alaska Volcano Observatory, the U.S. Air Force Weather Agency (AFWA) and Anchorage, Wellington and Washington VAACs. The japanese VAAC uses a modified release of the PUFF model. The European VAACs of London and Toulouse use respectively the Numerical Atmospheric dispersion Modeling Environment (NAME) and the MEDIA modeling environment.

A VATD model requires input data about the state of the atmosphere and is initialized by global or regional weather predictions. The availability of these weather data allows to provide wind fields, in order to estimate the dispersion of the ash clouds [6].

In this work I developed a high spatial and time weather service, using the main computing resource at University of Naples "Parthenope": the HPC-GPU cluster BlackJans (Figure 3.2).

The workflow composed by WRF-CALMET-CALPUF models was developed to ensure both the daily execution of the WRF model and to make available the needed weather dataset. If these data are available I can run the air quality dispersion model CALPUFF, which is modified to track the emissions of ash cloud due to volcanic eruption. The coupled model chain WRF-CALMET-CALPUFF is called ACT (Ash Cloud Tracking system).

According to the computing resources at our disposal, the ACT can be performed on all Italian territory, with a focus on southern Italy volcanoes, where there was a intense volcanic activity in recent years, with the emission of volcanic ash from Stromboli island and Mt. Etna; it also allows to predict what could happen in case of an eruption of Vesuvius.

2.2 Numerical weather prediction: an overview

Meteorology concerns the study of atmospheric dynamics and chemical-physical transformation of air masses. The air is continuously subjects to body forces, influenced by surface-atmosphere exchanges and meridional and zonal fluxes of energy and matter. Due to the physical and chemical complexities of atmospheric dynamics, we need to describe the circulation using a three-dimensional flow field extending over several scales, from a few meters to thousands of kilometers in horizontal and from tens of meters to the depth of the troposphere and over in the vertical [9, 10].

The dimensions of mesoscale can be defined for those atmospheric systems that have a horizontal extent large enough to allow the use of the hydrostatic approximation to describe the vertical pressure distribution. The hydrostatic approximation allows describing the state of the atmosphere: if the horizontal scales are larger than the vertical scales, using the atmospheric models with a horizontal grid sizes of the order of 100 km, the hydrostatic equation is useful.

However, in order to simulate smaller scale phenomena, which have vertical accelerations that are not negligible compared to buoyancy forces, such as storms or convective clouds, it is necessary to use the equations of motion without the hydrostatic approximation [11].

These scales of atmospheric phenomena, along with computing resources and economic cost limitations, define the domain and the grid sizes of mesoscale models [10].

Weather forecast is coupled with chemical and air quality models, determining the concentration levels of pollutants and their transport to other areas [9].

The foundation for any environmental model is a system of equations, descending from conservation principles. For mesoscale atmospheric models, these principles are:

1. conservation of mass;

2. conservation of heat;

3. conservation of motion;

4. conservation of water;

5. conservation of other gaseous and aerosol materials.

These equations are well known and they are widely developed within meteorological, marine and, adding the chemical trasformations, air quality numerical models, to be used into massive high performance computing (HPC) systems. These equations must be soveld simultaneously because they are inter-related. A system of nonlinear and partial differential equations (PDE) usually cannot be solved using analytic methods. According to [10] solving a PDE system requires numerical methods.

A numerical weather prediction (NWP) is an initial-value problem: giving an estimation of the atmospheric state, the model forecasts its evolution. The initial conditions and the boundary conditions for a PDE system are essential, in order to have a well-posed problem.

In mesoscale models, only the finite difference and the interpolation schemes are generally used, because they are easy to code into a computer; this technique involves the approximation of differential terms as in a Taylor series expansion [10]. Once selected, the equations are discretized and encoded into a program language. After identifying the numerical equations, it is necessary to define the domains and the grid structure to numerically solve the equations system. At last, we need to add boundary and initial conditions, as data input to equations system, required to provide unique solutions [10].

In meteorology the most used model is WRF (Weather Research and Forecast) to make weather forecast services and research.

In WRF model, to increase the spatial resolution and to describe the phenomena that occur at finer scales, nested domains are usually applied to achieve a high resolution output.

A domain can be decomposed using many kinds of grid schema: the grid schema of WRF is the Arakawa C, as showed in following Figure 2.1.

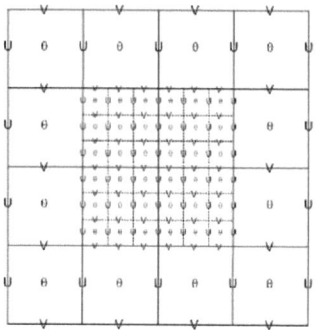

Figure 2.1: Nested domains using Arakawa C-grid

The u components are calculated on the left and on the right grid faces, and the v components on the center of the upper and of the lower grid faces.

The internal domains may be embedded simultaneously within a largest domain (parent) at coarser-resolution, or they run independently following a separate flow (Figure 2.2).

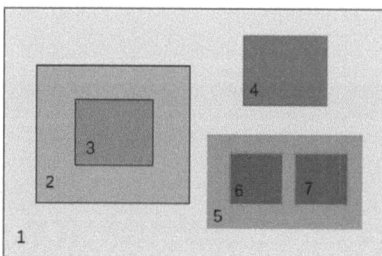

Figure 2.2. Possible relationships between WRF domains. The largest domain (1) have coarse resolution while the internal ones (2, 4, and 5) have middle resolution. The innermost domains (3,6 and 7) have fine resolution

The dependent variables within the system equations require initial values, before the integration process begins. For instance, values of wind speed, temperature, pressure perturbation are required at the startup of the simulation. The domains of mesoscale models are enclosed at four sides, therefore for these sides input values of the dependent variables are required at these border surfaces. These values, called boundary conditions, are required to integrate in time the approximate forms of the conservation relations.

The boundary conditions must be defined for top, lateral, and bottom limits of domains. It is important to select the limits of lateral and top boundary conditions because of constraints due to available computer resources, but expanding the domain horizontally is useful to minimize the effect of the lateral boundary. The top boundary conditions define the vertical stratification. These layers tend to generate circulations that have larger horizontal than vertical scales.

However, in an atmospheric model, the bottom boundary is the only one which has physical meaning and which increases its contribution to increase the spatial resolution of the grid. The bottom is a real boundary, and the transfer of physical properties, such as heat and moisture, across this interface [10].

2.3 The Weather Research and Forecast model

The Weather Research and Forecasting (WRF) model is a numerical weather prediction (NWP) and atmospheric simulation software, designed for both research and forecast services. The software development of WRF required the cooperation of many agencies efforts and a data assimilation system to improve the weather understanding and prediction [12].

As reported in [12] the WRF system consists of three modules for pre-processing, numerical solver and post-processing simulations. The relationships between these modules are shown in Figure 2.3

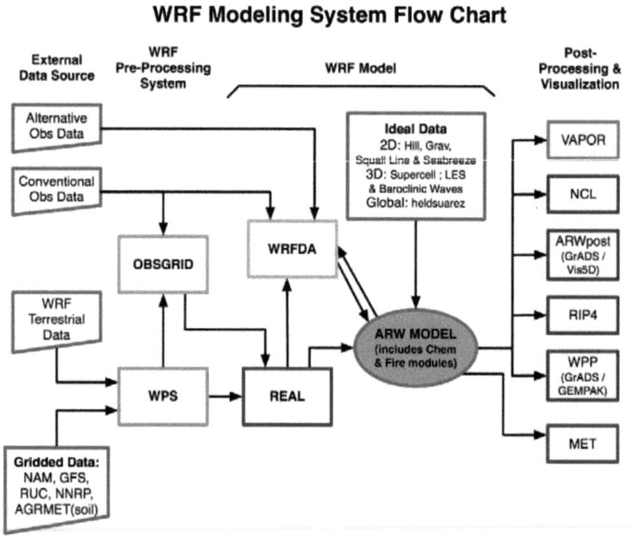

Figure 2.3: WRF modeling system: modules and their relationships (source: WRF User Web Page, updated 04/02/2010)

As shown in the block diagram in Figure 2.3, the WRF Modeling System consists of these main software modules:

- The WRF Preprocessing System (WPS);

- The WRF numerical solver, including a variational assimilation, chem and fire module (ARW);

- Post-processing & Visualization tools (ARWpost).

The WPS in Figure 2.4 is a package of tree modules: GEOGRID, UNGRIB and MET-GRID and its role includes:

1. defining simulation domains;

2. interpolating terrestrial data (such as terrain, landuse, and soil types) to the simulation domain;

3. degribbing and interpolating meteorological data from an external source to this simulation domain.

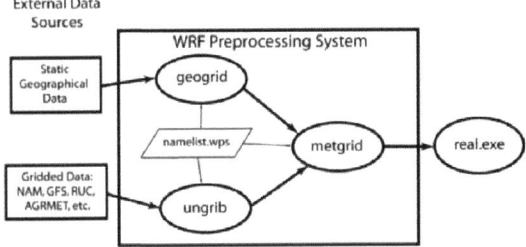

Figure 2.4: WPS modules and internal workflow

The UNGRIB program reads and extract the GRIB files, which contains the initialization data, and convert it in a "intermediate format". The GRIB files contain time-varying meteorological variables and they are typically obtained from global scale models distributed by U.S. Nationa Center for Enviromental Prediction (NCEP) and by the European Centre for Medium-Range Weather Forecasts (ECMWF).

The ARW is the equations dynamic solver, which is combined with other components of the WRF system; they are used together to obtain the forecast data about the evolution of the atmosphere.

In the Figure 2.5 are shown some example of modeling results published by weathere and marine center of University of Naples "Parthenope" (CCMMMA) in own web page.

The standard output of WPS, REAL, and WRF model is into NetCDF format (one of WRF I/O format) and it can be displayed by one or more multidimensional display tools. A commonly used display software, for multidimensional weather data, is GrADS (Grid Analysis and Display System), specifically developed for access, manipulation and visualization of earth science data.

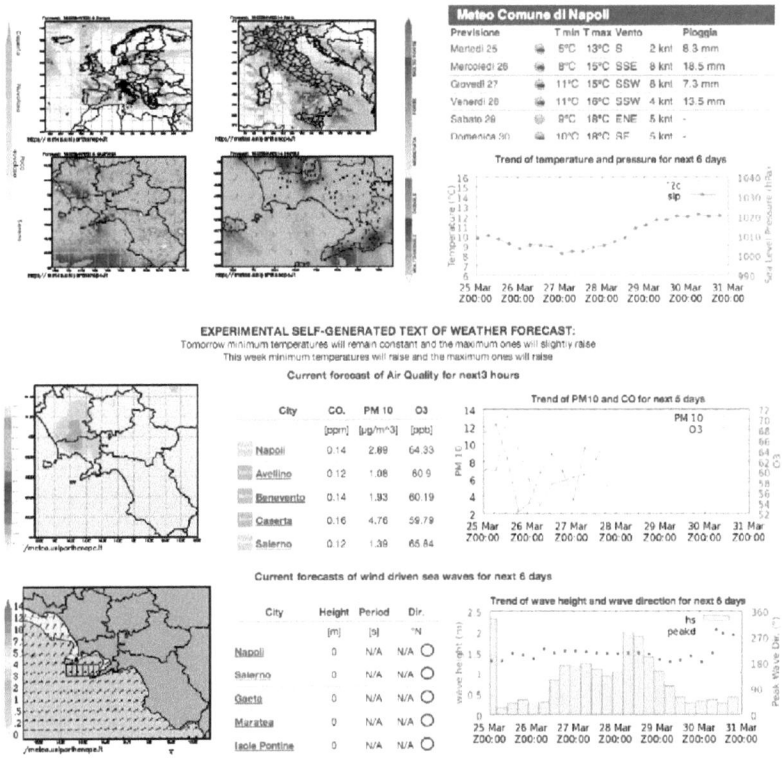

Figure 2.5: Detail of CCMMMA official web page with WRF, CHIMERE and WW3 output. In the picture we can see the maps of rain and cloud coverage, the distribution of PM10 and wave height. Also are shown the tables and time series of most important variables such as temperature and pressure, concentration of PM10, and peak wave direction.

2.4 The WRF initialization

The initialization datasets necessary to prepare a course-run of our WRF implementation are taken from NCEP; these data are obtained by an iterative data assimilation process using observed data. These data, which provide the state of the atmosphere, are collected using a global distributed atmospheric monitoring system (GOS) and are synchronized with a cycle of 6 hours, that is necessary to make an execution and share data produced by Global Forecast System (GFS) a numerical weather prediction system with an approximate horizontal resolution of 27km for the first 8 days and 35km from 192 to 384 hours (16 days).

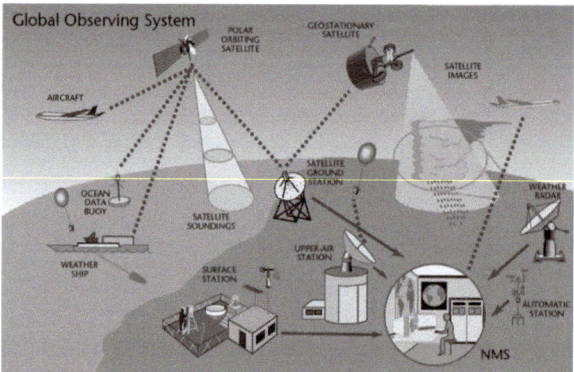

Figure 2.6: Global Observation System

The data generated by GFS are shared by NCEP are used to initilize the our implememntation of WRF.

Ensuring daily available WRF output data is a strategic target for an information weather service and not only for this one. We developed within of a distributed computing environment a shell scripting system that allows to obtain, every six hours (at 00:00, 06:00, 12:00 and 18:00 UTC), a forecast for the next following 144 hours. In our implememtation of WRF the nested domains has aspect ratio of 1:5 so the European, Italy and region domains have 25, 5 and 1 km of spatial mesh grid resolution respectively. At every course-run the WRF model was synchronized with the availability of NCEP initializing data and the forecast moves forward every six hour. These synchronized process must be managed with care because WRF offers multiple physics options that must be expertly combined. The combination of these options can change the behavior of WRF from fast and efficient, to slower and complex, with expensive computational times which implies a delay on the available of meteorological results at high resolution. The dataset produced by WRF allows the coupling with air quality models, such as CHIMERE, to follow the regional distribution of pollutants in the atmosphere, CALMET for grid diagnostic interpolation of wind fields to achieve an even higher resolution, typically from a few kilometers to hundreds meters and, last but not least, the CALPUFF system to forecast the Lagrangian trajectory of air masses and if it is applied to volcano eruptions, the dispersion of volcanic ash clouds. As shown in Figure 2.7, thanks at our implementation, the WRF has an execution time approximately of four hours, so we can grant free machine time for the execution of other models and applications.

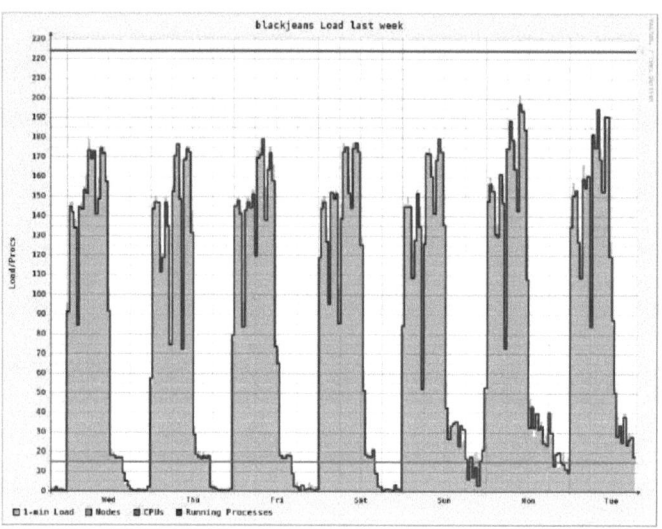

Figure 2.7: Report of work load of HPC-GPU Cluster BlackJeans of University of Naples "Parthenope"

The parameterization process of the chemical and physical properties of WRF model was made using the numerical schemes to approximate the solutions of the equations also according to our computing resources.

The most important parameterizations involve:

- Microphysics (mp_physics);

- Long- and Short-wave Radiation (ra_lw_physics/ra_sw_physics);

- Land Surface (sf_surface_physics);

- Planetary Boundary layer (bl_pbl_physics);

- Cumulus Parameterization (cu_physics).

In the next paragraphs we will explain in details their role within the WRF model. In order to obtain a high spatial resolution prediction, we need to make a correct tuning of these parameters, to not introduce errors those can affect the results.

2.4.1 Microphysics

The physical processes that lead to the formation, to the growth and to the precipitation of clouds are described by microphysics. The simulation process of phase transitions of water into an air mass can be represented to describe the degree of complexity of a simulation, like the growth to precipitation-sized liquid and ice particles [10].

In the WRF model it is possible to apply many microphysics schemes and the choice depends by the phenomena which are need to simulate. Considering our target, the spatial resolution in the innermost domain must have a mesh of 1 km, so we used the "New Thompson" scheme (mp_physics = 8) with ice and snow processes suitable for high resolution simulations [13].

2.4.2 Long- and Short-wave Radiation

The balance of incoming and outgoing thermal radiation on the Earth allows the distribution of heat on the planet from the equator to the poles.

The radiation balance allows the alternation of seasons and the control the weather conditions on the mesoscale. The Earth and the atmosphere emits the longwave radiation in the infrared band, while at the same time the shortwave radiation emitted from Sun reaches the Earth. The atmosphere has an important role in this balance, because the electromagnetic radiation can be absorbed, reflected, or transmitted [10].

In the WRF parametrization I choose to parameterize the longwave radiation using RRTM (Rapid Radiative Transfer Model) scheme, an accurate scheme using lookup tables for efficiency. The RRTM takes into account of multiple bands, trace gases, and microphysics species (ra_lw_physics = 1, 1, 1); these factors increase their contribution with the increasing of the spatial resolution and the total time of the simulation.

On the other hand, for shortwave radiation we used the Dudhia scheme, a simple downward integration allowing efficiently for clouds and clear-sky absorption and scattering (ra_sw_physics = 1, 1, 1) [13].

2.4.3 Land Surface

The representation of land surface as a bottom boundary requires different types of models than those required to properly represent the water interface. In contrast to water, the ground is opaque and does not readily overturn.

To represent land as a bottom surface, it is convenient to consider the bare soil separately from the vegetated ground. The former characterization is easier to simulate, and mesoscale models became increasingly more sophisticated in its representation. Vegetation effects, in contrast, are very complex.

Our WRF parametrization includes:

- The use of the VtableGFS in according to NCEP initializing data;

- The GEOGRID program which interpolates land using categories from USGS 24 category data (geog_data_res = '10m', '2m', '30s');

- 5-layer thermal diffusion: soil temperature scheme (sf_surface_physics = 1, 1, 1).

2.4.4 Planetary Boundary layer

The Planetary Boundary Layer (PBL) is the lowest layer of the troposphere and its behavior is directly influenced by planetary surface and friction.

In this layer the physical variables, such as flux velocity and as temperature, are character-ized by vertical mixing. Within the PBL the wind is affected by surface drag and it turns across the isobars. The higher layers of atmosphere are usually non-turbulent, or only intermittently turbulent [14]. The PBL schema in our WRF simulation was developed by Yonsei University with explicit entrainment layer and parabolic profile in unstable mixed layer (bl_pbl_physics = 1, 1, 1).

2.4.5 Cumulus Parameterization schemes

The cumulus parameterization schemes allow the WRF model to estimate the release of latent heat and convection motions within each grid box of the larger domain. These schemes are very important to estimate the convective rainfall and atmosphere heat and clouds phenomena.

Within the model the cumulus schema can theoretically release the latent heat in an un-stable environment, creating thunderstorms which can be simulated using the horizontal resolution into range from meters to tens of kilometers, and the domain can range from tens to thousands of kilometers.

However, the cumulus parameterization schemes are not developed to achieve fine spa-tial resolutions and they are useless for mesh greater than 4 km [13]. So in our WRF simulation we chose the Betts-Miller-Janjic scheme, with a well-mixed air column of the moisture profile (cu_physics = 2, 2, 0), but we turned off for the innermost domain at a resolution of 1 km.

2.4.6 Four-Dimensional Data Assimilation: Grid Nudging

The FDDA are a collection of different methods which can keep the simulations closer to analysis data during the course of integration. These techniques have a particular role to drive air quality chemistry models (e.g. CHIMERE); in our weather simulations we applied the grid-nudging method to force the model simulation towards a series of analyses grid-point by grid-point. The grid-nudging method is a three-dimensional analysis where the largest domain is kept towards time- and space-interpolated analyses, using a point-by-point relaxation term. The FDDA works on multiple domains in a nesting configuration; it requires multiple time-periods of each nudged domain as input analyses.

The method is implemented through an extra tendency term in the nudged variable's equations:

$$\frac{\partial \theta}{\partial t} = F\left(\theta\right) + G_\theta W_\theta \left(\hat{\theta}_0 - \theta\right) \tag{2.1}$$

where $F(\theta)$ represents the normal tendency terms due to physics, advection, etc., G_θ is a timescale controlling the nudging strength, and W_θ is an additional weight in time or space to limit the nudging as described more below, while $\hat{\theta}_0$ is the time- and space-interpolated analysis field value towards which the nudging relaxes the solution [12].

The grid-nudging works as a spring, keeping the simulation close to the initial conditions and preventing that variables fluctuate too much. The internal domains instead are free to evolve in response to external domain. This increases the accuracy of the simulation and it enables chemical models to evolve with greater regularity, containing oscillations due to boundary conditions.

2.5 An Overview of the CALPUFF Modeling System.

The availability of WRF weather prediction datasets at high spatial resolution is an important step to set-up an advanced early-warning system for volcanic ash tracking system. The wind fields at high resolution, with a grid scale at 1 km, have a fundamental role to initialize a dispersion model such as CALPUFF. CALPUFF is an advanced non-steady-state meteorological and air quality model. It is maintained by the model developers and distributed by the Atmospheric Studies Group (ASG), specialized in air quality model development, atmospheric boundary layer research, air quality regulatory consulting. The CALPUFF modeling system shown in Figure 2.8 is composed of three main components and a set of pre- and post-processing programs. The main components of the modeling system are CALMET (a diagnostic 3-dimensional meteorological model), CALPUFF (an air quality dispersion model), and CALPOST (a post-processing package). In addition to these components, there are many other modules that may be used to prepare geophysical data (land use and terrain) in several standard formats, as well as meteorological data (surface, upper air, precipitation, and buoy data); these modules have software interfaces to other models such as the WRF [15].

43

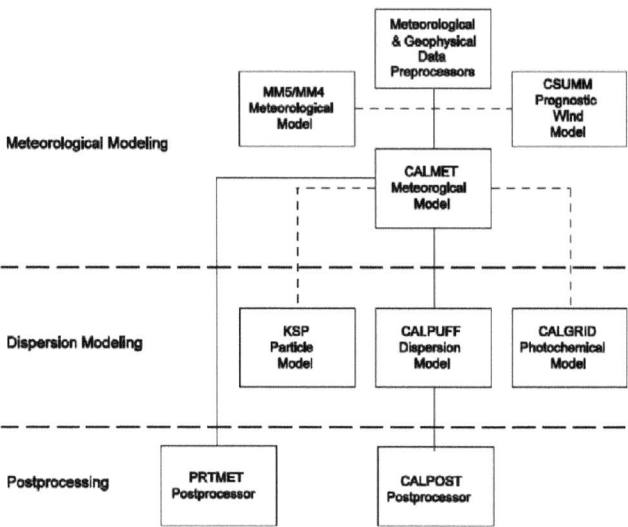

Figure 2.8: CALPUFF System overview (source: [3])

CALPUFF is a transport and dispersion model advecting "puff" of material emitted from the modeled source and simulate the dispersion and the transformations along a route using the interpolated wind fields generated by CALMET.

The module CALMET is a diagnostic wind model that allows to obtain hourly wind and temperature fields on a three-dimensional gridded domains, starting from the WRF output.

It is also a micro-meteorological model that allows to take into account the complex effects of terrain features on the wind circulation; terrain elevation, at high spatial resolution, within a range of hundred meters, plays an important role in the dispersion of an ash cloud.

The visibility of an aged volcanic ash layer will also be strongly hampered if it becomes embedded into the planetary boundary layer with high concentrations of particulate pollutants, which can easily be the case of industrialized regions [7].

3 Dispersion of Ash Cloud at high spatial and time resolution

Usually, weather dataset used to force an air quality dispersion model correspond to low time and spatial resolution. Data often consist of analysis available every $3 \div 6$ hours at a spatial resolution of $\approx 0.5° \times 0.5°$ in the horizontal directions [16]. This practice is very powerful for planning tests, validate theoretical and numerical models and develop the necessary auxiliary tools.

In this work it is suggested to improve this practice with an enhanced resolution: to ensure the spatial resolution needed for regional- and local-scale simulations, it is possible to couple meteorological models and defining local domains to achieve a finer mesh of hundred meters and a time resolution of few minutes. Moreover, at this spatial and time resolution the atmospheric phenomena on the local scale increase their contribution and it is necessary to take them into account to obtain an accurate dispersion of volcanic ash simulation.

A mesoscale weather forecasting system, e.g. WRF, requires large computing resources and a system to manage: workflow of data to ensure the coupling process with other numerical models; the right execution as single module.

In this work it was proposed a prototype wrapper for WRF-CALPUFF to enable the automation of a procedure and to link several modules, embedding dispersion of volcanic ash simulations in an early-warning network (models-instruments) and adding the possibility to use a standard visualization system, e.g. Google Earth. Users could be air traffic controllers, public and private agencies, so they can improve their own management capability of volcanic ash cloud risks.

In this chapter we described how to develop a software wrapper of the WRF and how it is possible to inizialize the CALPUFF system using a shell scripting framework, to allow the execution of an on-demand coupled workflow.

3.1 Weather forecast at high spatial and time resolution

The work made in the PhD project is part of the CCMMMA center of the University of Naples "Parthenope", whose aim is to become a regional reference center for advanced forecasting and environmental marine and air pollution monitoring. The center activities are focused on the implementation and the use of forecast modeling chain, producing daily information on the weather and sea state, on a regional scale. When weather data at high spatial and temporal resolution are obtained, they are used to initialize air quality (CHIMERE, CALMET-CALPUFF) and sea state (WW3, POM and SWAN) models.

The forecasting framework is supported by a monitoring network focused on the Gulf of Naples that consists of weather stations, weather radar, coastal radar and wavemeters.

In order to make a forecasting system for ash clouds using the CALPUFF model, it is necessary to make available high-resolution wind fields and to consolidate the behavior of the simulations to get the wind fields at high spatial and time resolution for those domains whose areas may be involved into volcanic eruptions with the emission of ash clouds.

The coupled models WRF-CALMET are optimized to produce daily forecasts, so our wind fields are updated every day, with a repetition cycle of six hours. The whole process, considering the computing resources at our disposal, takes about 4 hours to be successfully completed. A consistent part of this work was the development of shell scripts which can handle the data passing interfacing all modules of the WRF model. Our workflow ensures both automatically generation of daily weather forecasts for the Campania region and on-demand simulations for other region of southern Italy, was carried out.

In Figure 3.1 we show a workflow of the modeling system at the UniParthenope CCM-

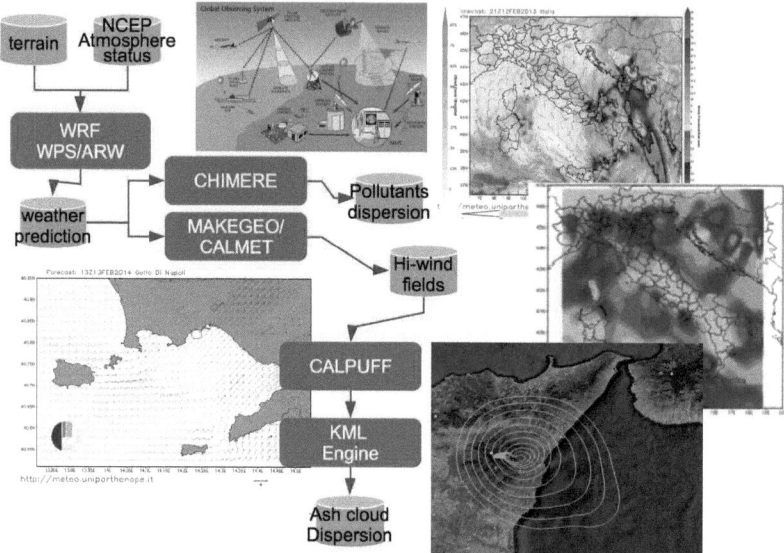

Figure 3.1: Overview of air quality modeling system used by UniParthenope CCMMMA center

MMA center.

The daily atmospheric and marine predictions were supported by the HPC-GPU cluster BlackJeans, designed to be the ideal computing environment and to have enough storage to allows collecting and sharing data (Figure 3.2).

In our system the daily WRF simulations are initialized with GFS data distributed by the NCEP. The WRF simulation is managed from a main shell script. The Actions performed by this shell script can be briefly summarized as follows:

1. Check NCEP data availability;

2. download NCEP dataset;

3. setup enviroment and model parameterization;

4. calculate the mesh grid, terrain and landuse for every nested domain (geogrid.exe);

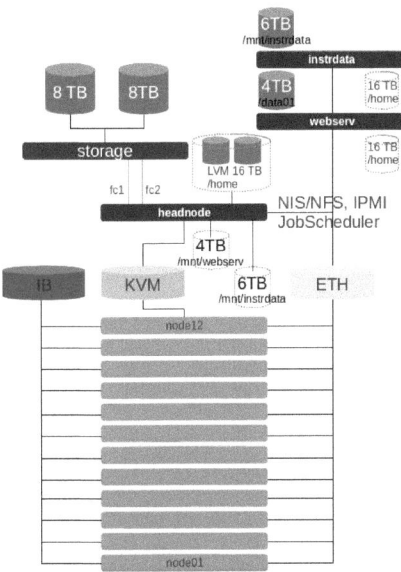

Figure 3.2: design schema of the HPC-GPU Blackjeans cluster. The cluster was composed by 12 computing node E7122, bi-processor Xeon SixCore X5650, 6 RAM modules DDR3-1333 4GB. Every computing nodes have a GPU NVIDIA Tesla M2050, one Network Attached Storage (NAS) fiber channel; one headnode E7122, 2 x Xeon 5506 QuadCore E5506, 6 RAM modules DDR3-1333 1GB; a switch Infiniband Qlogic 122000 36 slot QDR.

5. decompress and interpolate the NCEP dataset on the regional mesh (ungrib.exe and metgrid.exe);

6. interpolatate on vertical level and initialize the simulation (real.exe);

7. forecast the atmospheric dynamic for every nested domain (wrf.exe).

In Figure 3.3 we show a Google Earth grafical view of the innermost domains theorically available using our computing resource.

When a volcanic eruption occur, our system can use the most updated forecast data and increases the resolution, and hopefully that they are also the better reliability predictions available.

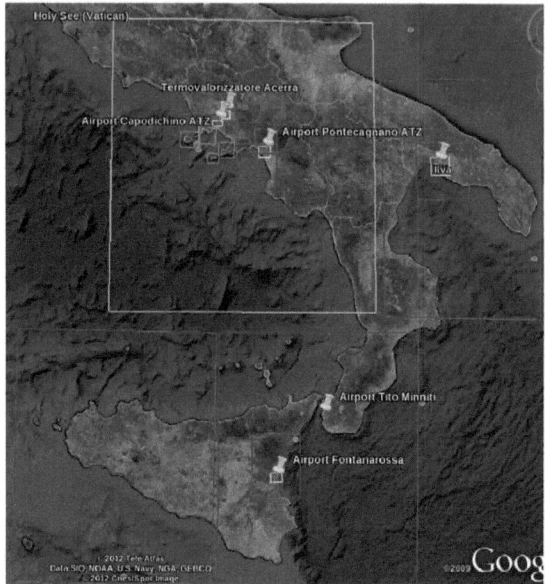

Figure 3.3: Available southern Italy nested domains from Google Earth.

3.2 The Ash Cloud Tracker:
WRF-CALMET-CALPUFF

When the WRF dataset is available it is possible to run the CALMET module to obtain the 3D wind fields at high spatial resolution for the area hit by a volcanic eruption with ashes emission. So it is possible to activate the on-demand CALPUFF module, whose output consists in the transport and the dispersion of chemicals and particles released into the atmosphere from active sources. With the CALPUFF module we can simulate the physical and chemical processes of substances along their track within the atmosphere, estimating the ground deposition flux.

We called our coupled modeling system ACT (Ash Clouds Traking) because it allows to predict the 3D distribution of volcanic ashes emitted by an eruptive event; it can be applied to improve air traffic management, optimize the management procedures for air space and coordinate ground operations. The block diagram in Figure 3.4 shows the workflow of the ACT system.

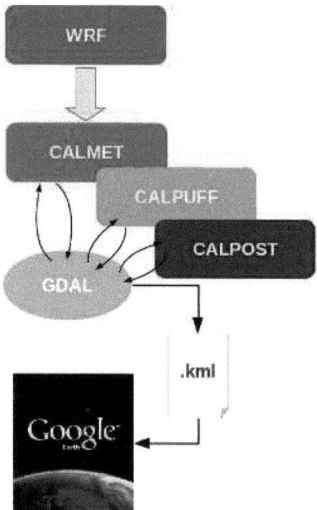

Figure 3.4: The ACT system: the process start from WRF; after the weather data are used as input by CALMET, we apply interpolation at high resolution with GEOTOPO terrain data over a domain with a grid with at resolution of few hundred meters (Figure 2.8). The wind fields are used to volcanic ash tracking with CALPUFF which allows modeling the features of a volcanic emission (temperature, exit velocity and radius of the vent). Finally, the CALPUFF output are convertend by KML Engine in kml format and visualized with Google Earth.

In order to simulate the physical characteristics of an eruptive event, the standard features of CALPUFF model were tuned taking into account the physical phenomena which guided the formation of the plume and its dissipation. The features were particle size distribution, temperature of the emitting fumes, position and radius of the vent, emission rate, mass fraction of water content.

3.3 ACT first applications: Stromboli Eruption on January 12th, 2013

During first year of PhD project we developed the shell script components which can handle the execution of WRF adding the CALMET module, giving particular attention at controlling the Input/Output data exchange from one model to another. The optimization of these procedure allowed to synchronize the execution clock-time to initialize the process and allows the prediction system to ensure the daily availability of high resolution data. The developing level reached at this point has made possible to apply the modeling chain WRF-CALMET to provide an advanced meteorological support for the two editions of the America's Cup World Series (2012 and 2013), those were carried out in the Bay of Naples (Italy).

In the second year, the work was dedicated to embed and manage the module CALPUFF, to simulate a volcanic ash emission on demand. When the most parts of development phases were concluded, it was possible to apply ACT for the first time, to provide a complete simulation of each components by the eruption of Stromboli on January 12th, 2013. According to the ufficial report of INGV, at 11.10 UTC the view of the Sciara was completely obscured for about 20 min by a dense cloud of steam load of particles that dated to the side gusts. The intense emission of vapor produced a cloud of charged particles policies of unusual magnitude that has developed over the summit of the volcano (INGV Report No. 03/2013). This experience provided important insights to improve the system and to give directions on how to handle input values passed to the CALPUFF module to properly simulate the transport of ash into the atmosphere.

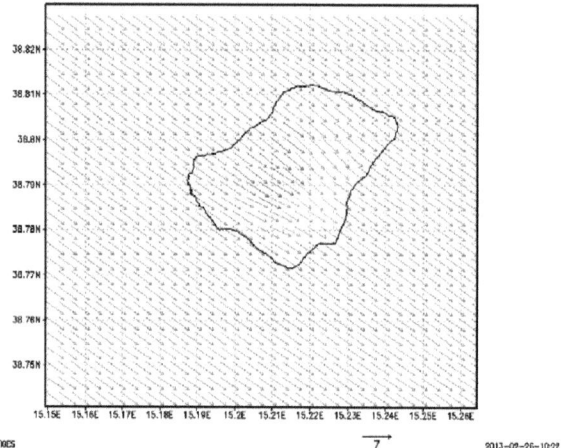

Figure 3.5: The interpolated wind fields of WRF-CALMET output on January 12th, 2013 at 12:00 a.m. UTC. This map was been visualized by GrADS (Grid Analisys Display System).

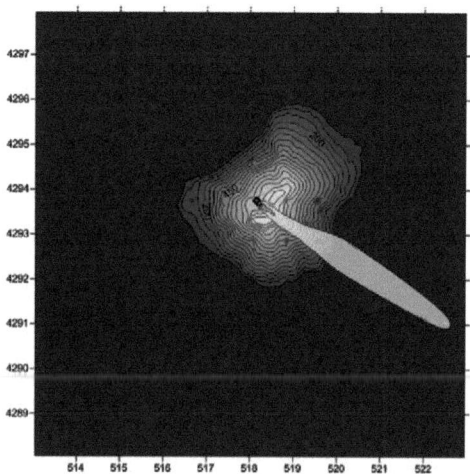

Figure 3.6: Ash cloud distribution by Stromboli eruption on January 12th, 2013 at 12:00 a.m. - UTC. This map can be displayed by common GIS software like QuantumGIS, ArcGIS or SURFER.

Moreover, our experience highlighted the need to provide a visualization tool that can be accessible to non-experts which can correctly interpret the results and that allows the users (air traffic controller, emergency staff of public and private agencies) to apply the appropriate response in case of volcanic ash risk.

On the third year, in order to consolidate the execution and to increase the reliability of data, we carried out sensitivity analyses of the simulations, comparing them to the values passed as input.

The on-demand data accessibility has been improved and the KML Engine module has been developed; it allowed to write data in KML format and enable the use of Google Earth as a visualization software.

KML (Keyhole Markup Language) is an eXtensible Markup Language (XML) schema based on open standard and maintained by the Open Geospatial Consortium (OGC).

The recent eruption of Mt. Etna began on December 14, 2013 and was of effusive and pyroclastic flow and it continued with the emission of volcanic ash also in the night of December 15th and 16th, 2013, with poor visibility conditions [17]. In this case we were able to initialize the ACT system and to make an assessment of the sensitivity of the simulations, comparing the input parameters with the official reports by the INGV.

We performed an execution of the WRF model for each day of eruption on three nested computational domains on Sicily region. The dataset of the innermost domain were centered to Mt Etna; the data were used as input for the wind fields at resolution of 500 meters to 200x200 grid points, centered to Mt Etna by the days of December 14th,15th and 16th, 2013.

The phases of WRF-CALMET, performed once for each days of the simulation, required about 4 hours and 30 minutes to be successful completed. When the wind data were available, we prepared a suite of 9 simulations, changing the input values such as the exit velocity of the flux of ashes and as the diameter of the vent. In total we obtained 27 simulations, nine for each day of eruption.

4 Ash Cloud Tracker system: high spatial resolution system

In the previous chapter we verified the possibility of using the WRF simulations at high spatial and time resolution as input data for the CALPUFF system (Figure 3.4). We saw how it is possible to obtain a dispersion of volcanic ash clouds, produced by volcanic eruption. In this chapter I show the distribution maps of volcanic ash obtained by our ACT. I applied the ACT system to an eruption of Mt. Etna and showed how, thanks to the system, it is possible to obtain the hourly dispersion of the ash cloud and also display the results using Google Earth.

4.1 Case Study: Eruption of Etna on December 14th,15th and 16th, 2013

The Mt. Etna is a stratovulcano situated on the east cost of Sicily; recently an intense volcanic activity was reported, characterized by effusive volcanic eruptions with ash emissions, that took many hardships for people. During these eruptions the nearby Catania-Fontanarossa Airport was repeatedly forced to close its runways, producing many inconveniences to air traffic. As reported by INGV [17] on December 14th, 2013 Mt. Etna takes a new eruption with initially discontinuous and then more regular ash emissions, until it determined the formation of a plume, that moved towards ESE (East-southeast). In the night between 14th and 15th, the emission of ash became more regular. Throughout the day of December 15th, the eruptive activity was accompanied by irregular ash emissions. Between the night of December 15th and 16th a clockwise rotation of the winds was recorded, that caused the dilution of the plume, producing dusts from SE (Southeast) to S (South) and SW (Southwest).

4.2 ACT applied to Eruption

As first step, I designed the WRF nested domains using the wrfDomainWizard, which allowed to prepare the configuration files of WRF: namelist.wps and namelist.input. In Figure 4.1(a) we can view an example of design phases.

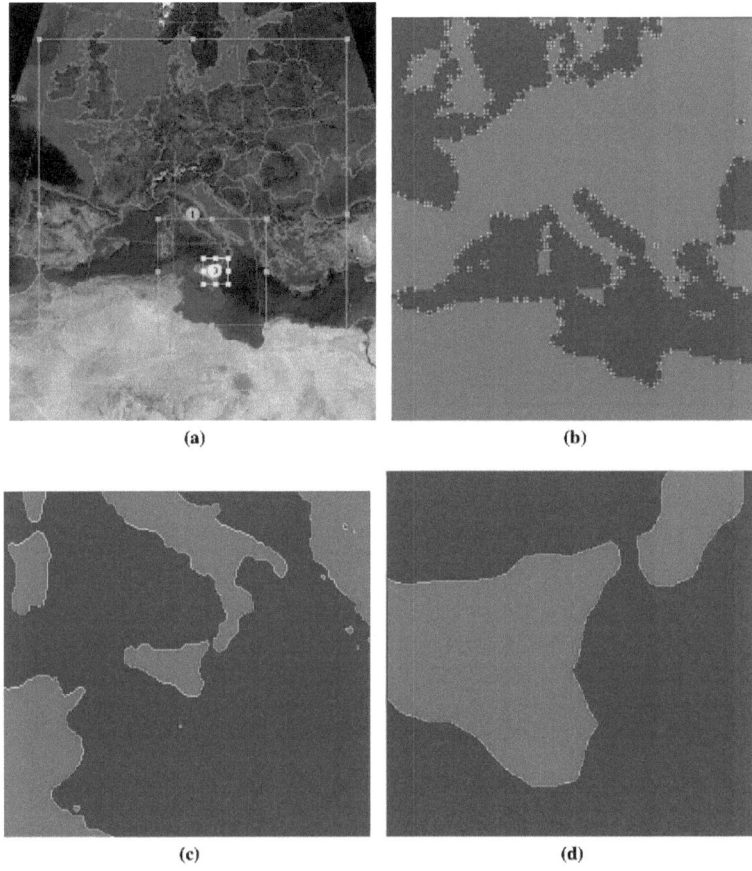

(a)   (b)

(c)   (d)

Figure 4.1: Nested domains of our WRF simulations. (a) Overview of wrfDomainWizard design, (b) the European region by mesh grid of 25 km, (c) southern Italy with a mesh of 5 km, (d) the Eastern Sicily centered on Mt. Etna with a mesh 1 km.

59

As second step, I defined a nesting ratio of 1:5 between domains, so the parent domain had a mesh of 25 km and covered the European region; the middle domain was centered on Mt. Etna, its mesh was 5 km and it covered Southern Italy; finally the innermost domain was also centered on Mt. Etna and it had a mesh of 1 km resolution and covered Eastern Sicily (Nested domains of our WRF simulations. (a) Overview of wrfDomainWizard design, (b) the European region by mesh grid of 25 km, (c) southern Italy with a mesh of 5 km, (d) the Eastern Sicily centered on Mt. Etna with a mesh 1 km.).

As shown in Chapter 2 the parameters, chosen for WRF simulations, were summarized in the following Physical schemes implemented for the WRF simulation on December 14th, 15th and 16th 2013.

Chemical and physical properties	WRF Schema
Microphisic	New Thompson
Long- and Short-wave	RRTM
Landuse and Terrain	VtableGFS, USG 24
PBL	Yonsei Univ.
Cumulus	BMJ
FDDA	Grid Nudging

Table 4.1: Physical schemes implemented for the WRF simulation on December 14th, 15th and 16th 2013

The configurations shown in Table 4.1 allowing to use the WRF to its maximum potential according our computing resources. On the third step, I compiled the WPS and ARW models of WRF models with PGI and MPICH2 library and we executed the WRF workflow with the HPC-GPU cluster Blackjeans in a distributed environment.

For each day of eruption on December 14th, 15 and 16th 2013, I initialized the WRF with NCEP dataset from 00:00 UTC. For each day we produced a forecast of 24 hours. Thanks to ACT system, the weather dataset were passed to the CALMET module to make a diagnostic interpolation on a finer mesh of 200x200 grid points and a resolution of 500 meters. The output of these steps were the hourly maps of wind fields at high resolution.

As fourth step, using wind fields obtained by CALMET it was possible initialize the CALPUFF model, preparing a set of simulations by varying the parameters that characterized the eruption. At the end of the CALPUFF phase I used the KML engine to convert the data in KML format. Then the maps in **??** were obtained using Google Earth as display system.

In the figure I show the concentration isolines of SO_2, the most important compound that we can find in volcanic ash emissions.

In Combination of ACT input parameters I show the combination of input values used to obtain the simulations made for this work. I modified the exit velocity of emission (EXITVEL in meter per second) and the radius vent (STACKD in meters). The values of EXITVEL were selected to be 20, 40 and 80 m/s and they were combined with the values of STACKD of 2, 5 and 7 m.

EXITVEL	STACKD
20	2
20	5
20	7
40	2
40	5
40	7
80	2
80	5
80	7

Table 4.2: Combination of ACT input parameters

The input values were chosen thanks to the integration of information about eruption published by INGV in official reports and thanks to data given by bibliography.

With this approach I could provide several hypothesis of scenario which could describe the eruptive phenomena and their impacts over the air traffic and over others human socio-economics activities.

4.3 Transport of ashes emitted by Mt. Etna

Our maps allow to find the areas impacted by eruptions, so the air traffic controllers and territorial agencies can plan out and can apply the most suitable safety strategy. The user, which has access to the ACT system, can read the scenario and can organize all the needed procedures to ensure flight safety in a timely manner.

In Figure 4.2 and in Figure 4.3 we can observe that the concentration isolines of SO_2 were closer to each other nearby the volcano vent: this fact indicated that the physical features of the eruption had more influence than the atmospheric status. The distance between isolines increased with distance from vent, the contribution of wind and of atmospheric status grow on the dispersion and on the downfall flux to the ground.

In Figure 4.2 the maps of distribution are compared with the wind map of 12:00 UTC and in Figure 4.3 at 18:00 UTC, both related to simulation on January 14th, 2013 initialized at 00:00 UTC.

Our weather forecast at 18:00 UTC showed how the decrease of wind energy corresponded to a decrease of distance between isolines and to an increase of ash downfall nearby the volcano vent. The urban areas and the runways of Catania-Fontanarossa airport were partially impacted by ashes. and the south airspace was free from particles.

The eastern Sicily is typically subject to a windy climate that carries the volcanic ash towards SE. The eruptive activity of the Mt. Etna often leads to the closure of the Catania-Fontanarossa airport, forcing to reschedule the flight routes in airspace of Catania. When Mt. Etna erupted on January 14th to 16th, 2013, at least 26 planes were forced to change their flight routes with rescheduling of air traffic management and consequent economic damage to airlines and air traffic management assessments.

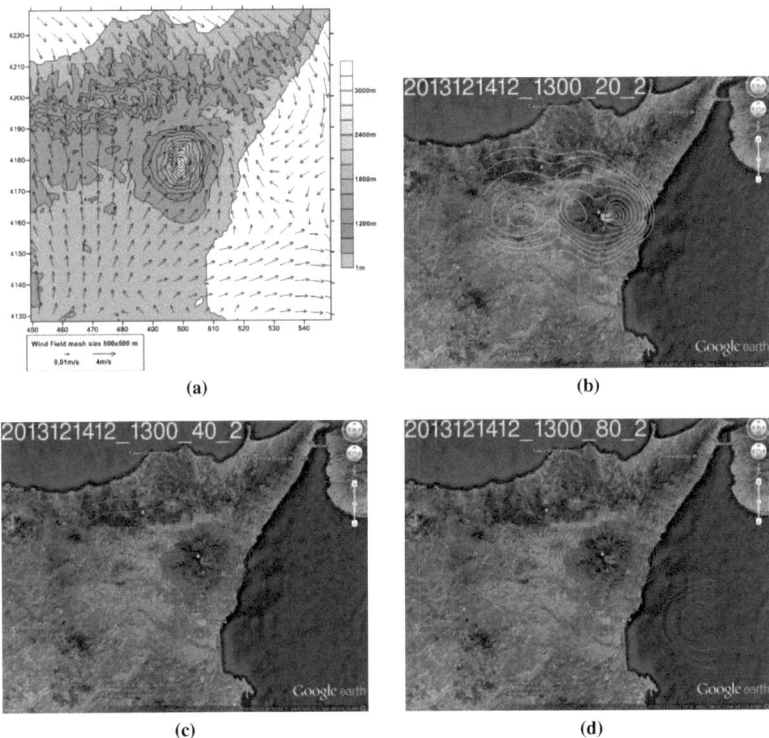

(a)

(b)

(c)

(d)

Figure 4.2: The wind map compared with SO_2 dispersion at 18:00 UTC on 14th Dec 2013 with Emission temperature 1300 K and radius vent 2 m. (a) Wind field at 12:00 UTC on 14th Dec 2013, (b) SO_2 dispersion flux to ground level with EXITVEL 20 m/s, (c) SO_2 dispersion flux to ground level with EXITVEL 40 m/s, (d) SO_2 dispersion flux to ground level with EXITVEL 80 m/s, the mass found in SE are formed by previous hours and is due to stationary condition according to wind circulation (a).

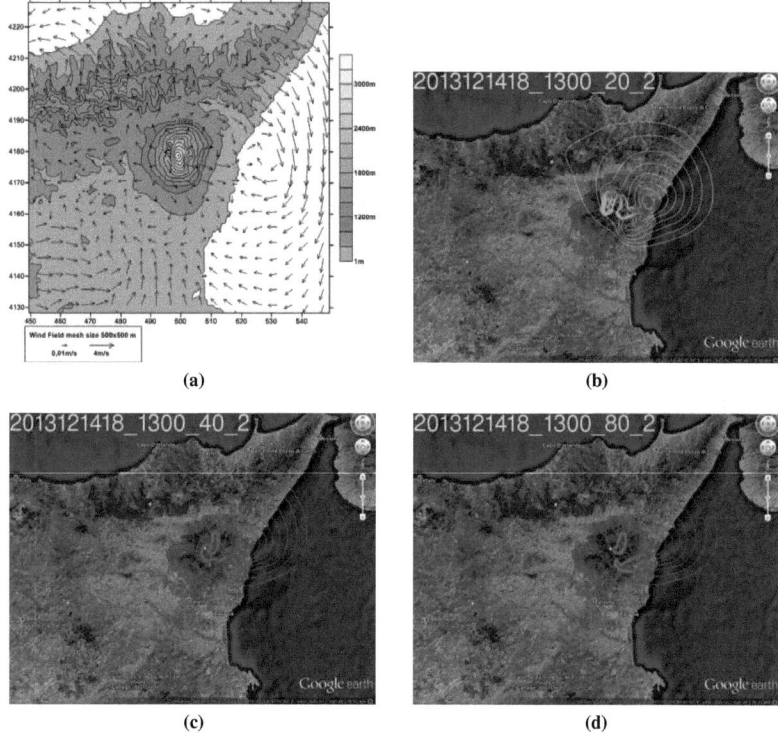

(a)   (b)

(c)   (d)

Figure 4.3: The wind map compared with SO_2 dispersion at 18:00 UTC on 14th Dec 2013 with Emission temperature 1300 K and radius vent 2 m. (a) wind field at 18:00 UTC on 14th Dec 2013, (b) SO_2 dispersion flux to ground level with EXITVEL 20 m/s, (c) SO_2 dispersion flux to ground level with EXITVEL 40 m/s, (d) SO_2 dispersion flux to ground level with EXITVEL 80 m/s

5 Conclusions and proposals for the future

In this thesis I demonstrated how it is possible to create a system of weather forecasting at high spatial and temporal resolution, using advanced computing resources, to obtain timely simulations of ash clouds dispersion. Moreover, in this work was shown how this system can be integrated in a environmental monitoring system.

In this chapter we discuss the use of our ACT system in real volcanic ash eruption and our considerations about the results. In addition we will advance some proposals for future ACT system improvement and its integration with other scientific methods.

5.1 Discussion of results

When a volcanic eruption occurs, large quantities of ashes may be emitted in atmosphere; the ashes are composed by particles of microscopic size and they have physical-chemical properties which make the airspace dangerous for flight.

In addition, when decrease the energy of eruption and meteorological condition becomes favor, the ashes fall to the ground. If the ashes cover the runways of airports there is a high danger for the landing/takeoff phases. The socio-economic impact of a volcanic eruption has very high costs for aviation business.

In southern Italy there are many areas affected by volcanic activity: in some of these areas there are airports which are always at risk of consequences by a volcanic eruption such as in case of Catania-Fontanarossa airport near Mt. Etna.

The weather forecast at high spatial and temporal resolution can be used to initialize tracking systems of volcanic ash on regional and local spatial grids in the range of hundreds of meters.

Our ACT system was composed by Eulerian models (WRF and CALMET) and a Lagrangian model (CALPUFF). It is handled by standard software tools developed and based on shell scripting. Also, the output of ACT was converted in KML format by our KML Engine module and the output data can be visualized by Google Earth.

To produce the weather dataset we need to ensure the data reliability, the data availability and also the computing resources, all of these factors are critical aspects in an environmental monitoring network, designed for early warning system.

The ACT system developed in this PhD project ensures the availability of weather data and the execution of volcanic ash dispersion models in case of emergency. The ACT system can be integrated and used by air traffic controllers and by public and private agencies which manage the territory.

In this work I applied the ACT system by performing sensitivity analyzes on the results obtained varying the input parameters which describe the eruption of Mt. Etna on mid-December 2013. For three days Etna erupted clouds of volcanic ash in the atmosphere even up to form a stable plume.

Our studies have shown that the system is able to predict the dispersion of volcanic ash and it allows to predict on a local scale the areas that will be affected by the ash-fall and the chemical compounds which composed the ash clouds.

Using ACT system we produced hourly forecast for 24 hours about the ash dispersion and the ash fall of SO_2. The ACT system can use weather data obtained with the WRF model which reaching a spatial scale of 1 km. The weather forecast data passed as input to the module CALMET, constituted by a 3D diagnostic interpolation, that allowed to obtain maps of the wind on a fine grid of 500 m. The maps of wind at high spatial and time resolution were used as input to the Lagrangian dispersion model CALPUFF, which allowed to predict the distribution of SO_2 in atmosphere and the flux to the ground of particles. The input parameters of the ACT system were modified according to the informations by INGV reports and we were able to produce large number of hypothetical scenarios.

In this thesis work some critical issues raised: necessity to obtain accurate informations about the eruption characteristics, the rate of ash emissions, the ashes exit temperature, the particles size distribution which made up the volcanic ash cloud. We tried to bypass these problems preparing more simulations of the same eruptive event, using the same weather conditions varying the parameters which describe these phenomena.

The velocity of emission (EXITVEL) of volcanic ash instead was varied within a given range and we were able to identify three representative values within which it is possible to include most of the eruptions of Etna. The diameter of the vent and the particle size distribution are the most difficult parameters to determine which make the most significant changes and we need to increase our confidence of the occurred eruptive event.

This target can be achieved if the ACT system will be added to a early-warning system and the calibrations will be made by integrating the observed data from the monitoring network composed by satellite, weather radar, webcam and lidar.

5.2 Proposals for the future

The computing resources have a very important role in the ACT system. If the computing resources are increase it is possible to reduce the time to execute the WRF model. Therefore, the meteorological data will be available in a shorter time and we can better schedule the computing resources. With a better scheduling of computing resources allows to generate a large number of on-demand instances of our ACT system, creating a larger number of simulations in shorter time.

The purposes of creation multiple instances of ACT system are following:

- obtain a bigger number of scenario;

- apply a data mining filter;

- compare the simulation data with real-data to identify patterns;

- select the best scenario of ash cloud dispersion.

The best scenario will selected to identify the patterns between the meteorological conditions and the distribution of ash, improving the ability of the users about the data interpretation.

The CALPUFF system consists of modules designed to be executed in serial computing environment. We will improve the model source code using standard libraries such as OpenMP and as MPICH, so we would create a parallel version of the software. This means being able to perform the full use of modern multi-core computing environments, both shared memory and distributed-memory.

If we can produce a large number of scenarios in a shorter time it would be possible identify the most likely scenarios and attribute them to a risk factor to timely identify the potential risk areas.

This approach would decrease the execution time of each ash cloud dispersion simulation and it would improve the integration of the ACT system within an early-warning system to volcanic ash risk assessment.

The data integration about ashes emitted by volcanic eruptions will improve the air quality forecasts on regional and local scale. Also, in next generation of air quality models, it is possible to develops a software module that allows to adding the contribution of volcanic ash emissions at runtime, this can improve the modeling of chemical reactions and allows to obtain a more accurate distribution and transport of particulate matter than now.

Bibliography

[1] T. M. Wilson, C. Stewart, V. Sword-Daniels, G. S. Leonard, D. M. Johnston, J. W. Cole, J. Wardman, G. Wilson, and S. T. Barnard, "Volcanic ash impacts on critical infrastructure," *Physics and Chemistry of the Earth, Parts A/B/C*, vol. 45, pp. 5–23, 2012.

[2] M. Guffanti, T. J. Casadevall, and K. E. Budding, *Encounters of aircraft with volcanic ash clouds: A compilation of known incidents, 1953-2009.* US Department of Interior, US Geological Survey, 2010.

[3] J. S. Scire, D. G. Strimaitis, and R. J. Yamartino, "A user's guide for the calpuff dispersion model," *Earth Tech, Inc*, vol. 521, pp. 1–521, 2000.

[4] "Flight safety and volcanic ash," ICAO - INTERNATIONAL CIVIL AVIATION ORGANIZATION, First Ediction Doc 9974/ ANB 487, 2012.

[5] M. Guffanti, G. C. Mayberry, T. J. Casadevall, and R. Wunderman, "Volcanic hazards to airports," *Natural hazards*, vol. 51, no. 2, pp. 287–302, 2009.

[6] P. Webley and L. Mastin, "Improved prediction and tracking of volcanic ash clouds," *Journal of Volcanology and Geothermal Research*, vol. 186, no. 1, pp. 1–9, 2009.

[7] B. Weinzierl, D. Sauer, A. Minikin, O. Reitebuch, F. Dahlkötter, B. Mayer, C. Emde, I. Tegen, J. Gasteiger, A. Petzold *et al.*, "On the visibility of airborne volcanic ash

and mineral dust from the pilot's perspective in flight," *Physics and Chemistry of the Earth, Parts A/B/C*, vol. 45, pp. 87–102, 2012.

[8] "Eruption scenario modelling and forecasting: the example of Mt. Etna, Italy," ICAO, Tech. Rep., 2011.

[9] J. H. Seinfeld and S. N. Pandis, *Atmospheric chemistry and physics: from air pollution to climate change.* John Wiley & Sons, 2012.

[10] R. A. Pielke Sr, *Mesoscale meteorological modeling*, 2002, vol. 78.

[11] E. Kalnay, *Atmospheric modeling, data assimilation, and predictability.* Cambridge university press, 2003.

[12] W. Shamarock, J. Klemp, J. Dudhia, D. Gill, D. Barker, M. Duda, X. Huang, W. Wang, and J. Powers, "A description of the advanced research wrf version 3," *NCAR technical note NCAR/TN/u2013475*, 2008.

[13] W. Wang, C. Bruyere, M. Duda *et al.*, "Wrf-arw v3: User's guide," *URL: http://www. mmm. ucar. edu/wrf/users (accessed: 11.01. 2013)*, 2010.

[14] (2014, January). [Online]. Available: http://en.wikipedia.org/wiki/Planetary_boundary_layer

[15] (2012). [Online]. Available: http://www.src.com/calpuff/calpuff1.htm

[16] S. Barsotti, D. Andronico, A. Neri, P. Del Carlo, P. Baxter, W. Aspinall, and T. Hincks, "Quantitative assessment of volcanic ash hazards for health and infrastructure at Mt. Etna (Italy) by numerical simulation," *Journal of Volcanology and Geothermal Research*, vol. 192, no. 1, pp. 85–96, 2010.

[17] "Bollettino settimanale sul monitoraggio vulcanico, geochimico e sismico del vulcano etna, 09/12/2013 - 15/12/2013," INGV, Tech. Rep. Rep 51/2013, 2013.

[18] B. Langmann, A. Folch, M. Hensch, and V. Matthias, "Volcanic ash over europe

during the eruption of eyjafjallajökull on iceland, april–may 2010," *Atmospheric Environment*, vol. 48, pp. 1–8, 2012.

[19] P. Webley, J. Dehn, J. Lovick, K. Dean, J. Bailey, and L. Valcic, "Near-real-time volcanic ash cloud detection: Experiences from the alaska volcano observatory," *Journal of Volcanology and Geothermal Research*, vol. 186, no. 1, pp. 79–90, 2009.

[20] L. Mastin, M. Guffanti, R. Servranckx, P. Webley, S. Barsotti, K. Dean, A. Durant, J. Ewert, A. Neri, W. Rose *et al.*, "A multidisciplinary effort to assign realistic source parameters to models of volcanic ash-cloud transport and dispersion during eruptions," *Journal of Volcanology and Geothermal Research*, vol. 186, no. 1, pp. 10–21, 2009.

[21] P. Webley and L. Mastin, "Improved prediction and tracking of volcanic ash clouds," *Journal of Volcanology and Geothermal Research*, vol. 186, no. 1, pp. 1–9, 2009.

[22] S. A. Carn, A. J. Krueger, N. A. Krotkov, K. Yang, and K. Evans, "Tracking volcanic sulfur dioxide clouds for aviation hazard mitigation," *Natural hazards*, vol. 51, no. 2, pp. 325–343, 2009.

[23] A. Prata and J. Kerkmann, "Simultaneous retrieval of volcanic ash and so2 using msg-seviri measurements," *Geophysical Research Letters*, vol. 34, no. 5, 2007.

[24] R. J. Spence, P. J. Baxter, and G. Zuccaro, "Building vulnerability and human casualty estimation for a pyroclastic flow: a model and its application to vesuvius," *Journal of Volcanology and Geothermal Research*, vol. 133, no. 1, pp. 321–343, 2004.

[25] D. R. Stauffer and J.-W. BAO, "Optimal determination of nudging coefficients using the adjoint equations," *Tellus A*, vol. 45, no. 5, pp. 358–369, 1993.

[26] R. Borge, V. Alexandrov, J. José del Vas, J. Lumbreras, and E. Rodríguez, "A com-

prehensive sensitivity analysis of the wrf model for air quality applications over the iberian peninsula," *Atmospheric Environment*, vol. 42, no. 37, pp. 8560–8574, 2008.

[27] P. Daniele, L. Lirer, P. Petrosino, N. Spinelli, and R. Peterson, "Applications of the puff model to forecasts of volcanic clouds dispersal from etna and vesuvio," *Computers & Geosciences*, vol. 35, no. 5, pp. 1035–1049, 2009.

[28] S. Barsotti, A. Neri, and J. Scire, "The vol-calpuff model for atmospheric ash dispersal: 1. approach and physical formulation," *Journal of Geophysical Research: Solid Earth (1978–2012)*, vol. 113, no. B3, 2008.

[29] S. Barsotti, L. Nannipieri, and A. Neri, "MAFALDA: An early warning modeling tool to forecast volcanic ash dispersal and deposition," *Geochemistry, Geophysics, Geosystems*, vol. 9, no. 12, 2008.

[30] S. Barsotti and A. Neri, "The vol-calpuff model for atmospheric ash dispersal: 2. application to the weak mount etna plume of july 2001," *Journal of Geophysical Research: Solid Earth (1978–2012)*, vol. 113, no. B3, 2008.

[31] B. J. Stunder, J. L. Heffter, and R. R. Draxler, "Airborne volcanic ash forecast area reliability," *Weather & Forecasting*, vol. 22, no. 5, 2007.

[32] US-EPA, *Documentation of the Evaluation of Calpuff and Other Long Range Transport Models Using Tracer Field Experiment Data.* BiblioGov, 2013.

[33] C. Searcy, K. Dean, and W. Stringer, "Puff: A high-resolution volcanic ash tracking model," *Journal of Volcanology and Geothermal Research*, vol. 80, no. 1, pp. 1–16, 1998.

[34] J. S. Scire, F. R. Robe, M. E. Fernau, and R. J. Yamartino, "A user's guide for the calmet meteorological model," *Earth Tech, USA*, vol. 37, 2000.

[35] V. Matthias, A. Aulinger, J. Bieser, J. Cuesta, B. Geyer, B. Langmann, I. Serikov, I. Mattis, A. Minikin, L. Mona *et al.*, "The ash dispersion over europe during the

eyjafjallajökull eruption–comparison of cmaq simulations to remote sensing and air-borne in-situ observations," *Atmospheric Environment*, vol. 48, pp. 184–194, 2012.

[36] A. Papayannis, R. Mamouri, V. Amiridis, E. Giannakaki, I. Veselovskii, P. Kokkalis, G. Tsaknakis, D. Balis, N. Kristiansen, A. Stohl *et al.*, "Optical properties and vertical extension of aged ash layers over the eastern mediterranean as observed by raman lidars during the eyjafjallajökull eruption in may 2010," *Atmospheric Environment*, vol. 48, pp. 56–65, 2012.

[37] D. C. Carslaw, M. L. Williams, and B. Barratt, "A short-term intervention study-impact of airport closure due to the eruption of eyjafjallajökull on near-field air quality," *Atmospheric Environment*, vol. 54, pp. 328–336, 2012.

[38] M. Bursik, S. Kobs, A. Burns, O. Braitseva, L. Bazanova, I. Melekestsev, A. Kurbatov, and D. Pieri, "Volcanic plumes and wind: Jetstream interaction examples and implications for air traffic," *Journal of Volcanology and Geothermal Research*, vol. 186, no. 1, pp. 60–67, 2009.

[39] "Volcanic ash contingency plan eur and nat regions," ICAO, Tech. Rep. EUR Doc 019 NAT Doc 006, Part II, 2010.

[40] R. De Laura, M. Robinson, R. Todd, and K. MacKenzie, "Evaluation of weather impact models in departure management decision support: operational performance of the route availability planning tool (rapt) prototype," in *13th Conference on Aviation, Range, and Aerospace Meteorology, AMS, New Orleans, LA*, 2008.

[41] M. M. Wolfson and D. A. Clark, "Advanced aviation weather forecasts," *Lincoln Laboratory Journal*, vol. 16, no. 1, p. 31, 2006.

[42] M. Guffanti, G. C. Mayberry, T. J. Casadevall, and R. Wunderman, *Compilation of Disruptions to Airports by Volcanic Activity (version 1.0, 1944-2006)*. US Geological Survey, 2007.

Printed by Books on Demand GmbH, Norderstedt / Germany